HSP Pennsylvania Science

PSSA Preparation for Science

Grade 3

Harcourt
SCHOOL PUBLISHERS

Visit The Learning Site!
www.harcourtschool.com

Copyright © by Harcourt, Inc.

All rights reserved. No part of this publication may be reproduced or transmitted in any form or by any means, electronic or mechanical, including photocopy, recording, or any information storage and retrieval system, without permission in writing from the publisher.

Permission is hereby granted to individuals using the corresponding student's textbook or kit as the major vehicle for regular classroom instruction to photocopy Copying Masters from this publication in classroom quantities for instructional use and not for resale. Requests for information on other matters regarding duplication of this work should be addressed to School Permissions and Copyrights, Harcourt, Inc., 6277 Sea Harbor Drive, Orlando, Florida 32887-6777. Fax: 407-345-2418.

HARCOURT and the Harcourt Logo are trademarks of Harcourt, Inc., registered in the United States of America and/or other jurisdictions.

Printed in the United States of America

ISBN-13: 978-0-15-374749-6
ISBN-10: 0-15-374749-8

2 3 4 5 6 7 8 9 10   862   16 15 14 13 12 11 10 09 08

If you have received these materials as examination copies free of charge, Harcourt School Publishers retains title to the materials and they may not be resold. Resale of examination copies is strictly prohibited and is illegal.

Possession of this publication in print format does not entitle users to convert this publication, or any portion of it, into electronic format.

Contents

Learning About the Pennsylvania System of School Assessment
   (PSSA) Examination iv

Practice Sets
   Getting Ready for Science through Chapter 16 1–68

Practice Tests

   Test 1 69

   Test 2 87

   Test 3 105

Learning About the Pennsylvania System of School Assessment (PSSA) Examination

The next two pages give examples of the types of items found on the science portion of the Pennsylvania System of School Assessment (PSSA) Examination.

About Multiple-Choice Items

Many of the items on the Pennsylvania System of School Assessment (PSSA) Examination are multiple-choice. Each multiple-choice item has four answer choices. The tips that follow will help you answer these questions.

1. Read the question carefully. Restate the question in your own words.

2. Watch for key words such as *best, most, least,* or *except*.

3. The question might include tables, graphs, diagrams, or pictures. Study these carefully before choosing an answer.

4. Find the best answer for the question. Fill in the answer bubble for that answer. Do not make any stray marks around answer spaces.

Use the illustration to answer question 1.

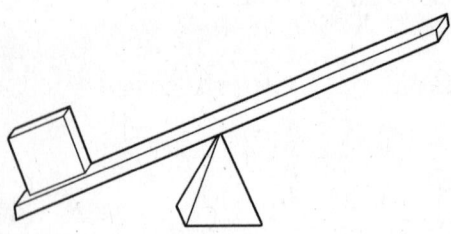

1. What kind of simple machine is shown in the illustration?

    (A) wheel-and-axle

    (B) inclined plane

    (C) pulley

    (D) lever

About Constructed-Response (Open-Ended) Items

For some items, you must write a brief answer to explain a science concept or to apply a science process skill. To receive the highest score answers should

- be complete.
- show understanding of the science content and processes.
- be accurate.
- communicate the ideas clearly.

---

1. The three pictures can be used to make a food chain.

   A. Identify the order the pictures should be put in to make a food chain.

   _____

   _____

   B. How is energy transferred between the three living things in the food chain you made?

   _____

   _____

   _____

   _____

About Practice Sets

Each practice set consists of three multiple-choice items and 2 two-point open-ended items. Some items have Tips, which give clues about how to answer the items. An item may have a graph, table, picture, or diagram. Study these carefully before answering the items.

Name _____

Date _____

Getting Ready for Science
Practice Set

1. Which tool is used to pick up and release small amounts of liquid?

Tip
Look carefully at each tool. Reread the question to look for key words.

Ⓐ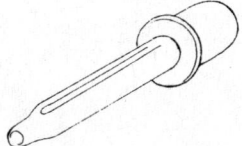

Ⓑ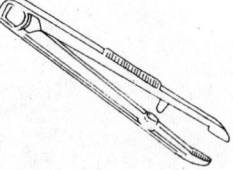

Ⓒ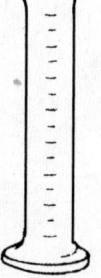

Ⓓ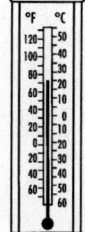

Name _____ Date _____

2. Which tool is used to measure a solid object that is curved?

   A thermometer

   B ruler

   C measuring tape

   D measuring cup

> **Tip**
> Remember what you are measuring. Think about how you would measure it.

3. Which inquiry skill would you use to find out how two objects are alike?

   A classify/order

   B compare

   C formulate

   D identify and control variables

Getting Ready for Science Practice Set

Name _____ Date _____

4. *Ask a question* is usually the first step in the scientific method. Identify at least two more steps of the scientific method.

Tip
Suppose that you are carrying out an investigation. Think about the steps you would take and the order you would follow.

_____

_____

_____

_____

(page 3 of 4) **Getting Ready for Science** Practice Set

Name _____ Date _____

Use the illustration below to answer question 5.

5. You want to measure a soccer ball like the one shown here.

   Tip
   Think about the different ways to measure things.

   A. What property of the soccer ball could you measure?

   _____
   _____
   _____

   B. What tool would you use to make this measurement?

   _____
   _____
   _____
   _____

4 Getting Ready for Science **Practice Set**

Name _____

Date _____

Chapter 1 Practice Set

Use the diagram below to answer question 1.

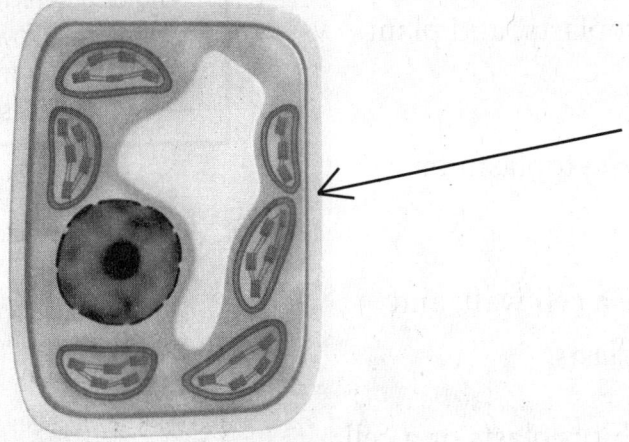

1. To what part of the plant cell is the arrow pointing?

    Ⓐ cell wall

    Ⓑ nucleus

    Ⓒ vacuole

    Ⓓ cytoplasm

Tip
Remember that the diagram shows a plant cell. Recall what the outer layer of the cell is called.

Name _____ Date _____

2. Which are two ways that animal cells differ from plant cells?

   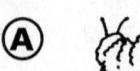 Animals cells have chloroplasts, and plant cells have a nucleus.

   Ⓑ Animal cells do not have cytoplasm or a nucleus.

   Ⓒ Animal cells do not have a cell wall, and they do not have chloroplasts.

   Ⓓ Plant cells do not have chloroplasts or a cell wall.

   > **Tip**
   > Look for the answer choice in which both examples tell how animal cells differ from plant cells.

3. Which shows the first stage in the metamorphosis of a butterfly?

   Ⓐ   Ⓒ

       Ⓓ

6 Chapter 1 Practice Set

(page 2 of 4)

Name _____ Date _____

4. What is the name of the change that a grasshopper goes through during its life cycle? Describe how the grasshopper changes as it grows.

> **Tip**
> Think about how the grasshopper looks at each stage of its life cycle.

_____

_____

_____

_____

Name _____ Date _____

5. Ross has freckles, brown eyes, and straight brown hair. His mom has blue eyes and brown hair, and his dad has brown eyes and blond hair.

> **Tip**
> Compare the traits Ross has to the traits his parents have.

A. What traits does Ross have from his mother? What traits does he have from his father?

_____

_____

B. Ross learned to walk by age 1 and to talk by age 2. Did Ross inherit these abilities from his mother or father? Explain your reasoning.

_____

_____

_____

_____

Chapter 1 Practice Set

Name _____

Date _____

Chapter 2 Practice Set

1. Which part of a plant takes in water and nutrients from the soil?

Tip
Identify the part of the plant that grows underground.

Ⓐ

Ⓑ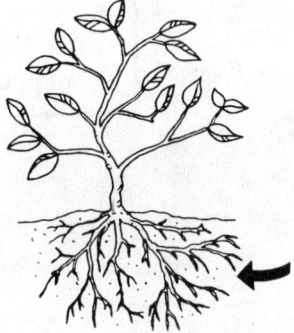

Ⓒ

Ⓓ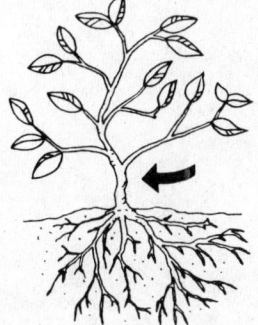

Name _____ Date _____

2. How are deciduous plants and evergreen plants alike?

   (A) Both have large, flat leaves.

   (B) Both lose their leaves each year.

   (C) Both reproduce using spores.

   (D) Both have leaves that make food.

---

3. Chlorophyll gives leaves their green color. How does chlorophyll help plants make food?

   (A) Chlorophyll helps plants take in nutrients.

   (B) Chlorophyll helps plants turn oxygen into sugar.

   (C) Chlorophyll helps plants take in carbon dioxide.

   (D) Chlorophyll helps plants collect and use the sun's energy.

   > **Tip**
   > Think about the process of photosynthesis.

Chapter 2 Practice Set

Name _____ Date _____

4. Scientists group similar plants together. What is one way that scientists classify plants into two large groups?

> **Tip**
> Think of characteristics that certain plants share.

_____

_____

_____

_____

Name _____ Date _____

5. Tico is doing an experiment with plants. He plants three small plants in three separate paper cups. He uses the same amount and type of soil, and he places all three plants on a sunny windowsill. He does not water Plant A. He waters Plant B until the soil is slightly damp. He waters Plant C until there is water standing on top of the soil. He does the same thing every day for two weeks.

> **Tip**
> Identify ways to measure plant growth. Think of how the plants will differ after two weeks.

A. What data about the plants should Tico record?

_____

_____

_____

_____

B. What do you predict will happen to the plants after two weeks?

_____

_____

_____

_____

_____

_____

_____

Name _____

Date _____

Chapter 3 Practice Set

Use the illustrations below to answer question 1.

1. The illustrations show an amphibian and a reptile. How do amphibians and reptiles differ?

   Ⓐ Amphibians have live young, and reptiles hatch from eggs.

   Ⓑ Amphibians live only on land, and reptiles live only in the water.

   Ⓒ Amphibians do not have backbones, and reptiles do have backbones.

   Ⓓ Amphibians have moist skin, and reptiles have dry skin.

Tip
Recall the characteristics of amphibians and the characteristics of reptiles.

Name _____ Date _____

2. Which of these are basic needs of all animals?

   Ⓐ food, water, shelter, and color

   Ⓑ shelter, water, food, and oxygen

   Ⓒ water, food, shelter, and love

   Ⓓ love, food, oxygen, and color

3. What is a function of the scales on a fish?

   Ⓐ protection

   Ⓑ eating

   Ⓒ giving birth to young

   Ⓓ breathing

   Tip
   Think about where scales are found on fish. What do fish scales feel like?

Chapter 3 Practice Set

Use the illustrations below to answer question 4.

4. Both animals in the picture live in the ocean and are vertebrates, but these animals have other characteristics that are very different. Identify two ways these animals differ.

Tip
Make a Venn diagram to compare and contrast these animals.

_____
_____
_____
_____

Name _____ Date _____

5. Scientists classify animals into groups.

   A. What are two important ways in which scientists classify animals?

   Tip
   Remember that you would be looking for clues you could observe.

   _____
   _____
   _____
   _____

   B. If you found an animal you had never seen before, what would you look for in order to classify it?

   _____
   _____
   _____
   _____
   _____
   _____

Name _____

Date _____

Chapter 4 Practice Set

1. What two changes happen to an animal during hibernation?

    (A) Its body temperature drops, and its heartbeat slows.

    (B) It uses a lot of energy, and its heartbeat speeds up.

    (C) Its body temperature rises, and its heartbeat speeds up.

    (D) Its body temperature drops, and it uses a lot of energy.

> **Tip**
> Look for the answer choice in which both changes happen during hibernation.

Name _____ Date _____

Use the picture below to answer question 2.

2. Which ecosystem does the picture show?

   Ⓐ grassland

   Ⓑ desert

   Ⓒ tundra

   Ⓓ swamp

---

3. The viceroy butterfly looks like the monarch butterfly, which tastes bad to birds. Birds often mistake the viceroy for a monarch and leave it alone. What type of adaptation does the viceroy butterfly show?

   Ⓐ hibernation

   Ⓑ instinct

   Ⓒ migration

   Ⓓ mimicry

Tip
Look for key words to help you identify the adaptation.

Name _____ Date _____

4. How do the shallow roots of a cactus help it survive in a desert ecosystem?

> **Tip**
> What is the main characteristic of a desert ecosystem?

_____

_____

_____

_____

Name _____ Date _____

5. Plans are being made to clear a large part of a deciduous forest to build houses.

> **Tip**
> How will these changes affect the habitats of plants and animals?

A. What resources of the forest will people use to build houses?

_____

_____

B. What will happen to the forest plants and animals when the houses are built?

_____

_____

_____

_____

_____

Chapter 4 Practice Set

Name _____

Date _____

Chapter 5 Practice Set

1. Wolves hunt rabbits and mice for food. Which are the predators in this example?

    Ⓐ the mice

    Ⓑ the rabbits

    Ⓒ the wolves

    Ⓓ the food

2. Which of these names only producers?

    Ⓐ insects and grass

    Ⓑ grass and trees

    Ⓒ dirt and deer

    Ⓓ trees and frogs

Tip
Look for the answer choice in which both are producers.

Name _____ Date _____

3. Which of the following do herbivores eat?

   Ⓐ carrots and rabbits

   Ⓑ berries and leaves

   Ⓒ mice and insects

   Ⓓ berries and birds

> **Tip**
> Eliminate answer choices that name foods herbivores do not eat.

Name _____ Date _____

4. Several owls live in a forest. The owls prey on mice. Hawks also prey on mice. What will most likely happen if several hawks move to the forest where the owls live?

> **Tip**
> Think about the food chain and what might happen to animals that depend on the same food source.

_____

_____

_____

_____

_____

_____

Name _____ Date _____

5. These three pictures can be used to make a food chain.

> **Tip**
> Think about how energy flows between living things.

A. Identify the order the pictures should be put in to make a food chain.

_____

_____

B. How is energy transferred between the three living organisms in the food chain you made?

_____

_____

_____

_____

_____

24 Chapter 5 Practice Set

Name _____

Date _____

Chapter 6 Practice Set

1. In which kind of rocks do most fossils form?

    Ⓐ igneous rock

    Ⓑ metamorphic rock

    Ⓒ sedimentary rock

    Ⓓ volcanic rock

Tip
Recall how the different types of rocks form.

Use the illustration below to answer question 2.

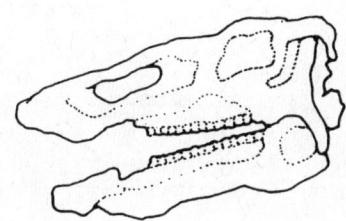

2. Scientists found the fossil of an animal, as shown in the illustration. Notice that the teeth are flat, not sharp. What can the scientists infer about the animal?

    Ⓐ The animal ate meat.

    Ⓑ The animal was small.

    Ⓒ The animal was large.

    Ⓓ The animal ate plants.

Name _____ Date _____

3. Igneous rock is rock that was once melted and then cooled and hardened. What can cause igneous rock to form?

> **Tip**
> Recall the definition of *igneous rock*.

Ⓐ mountains eroding

Ⓑ volcanoes erupting

Ⓒ earthquakes

Ⓓ layers of dirt being squeezed together

Name _____ Date _____

Use the chart below to answer question 4.

Mohs Hardness Scale	
Minerals	**Hardness**
Talc	1
Gypsum	2
Calcite	3
Fluorite	4
Apatite	5
Feldspar	6
Quartz	7
Topaz	8
Corundum	9
Diamond	10

Tip
Recall the meaning of the numbers on the Mohs scale.

4. Look at the Mohs Hardness Scale. Rank the following minerals from softest to hardest: apatite, gypsum, topaz, and quartz.

_____

_____

Name _____ Date _____

5. Amanda was walking on the island of Hawaii. She was told that there are several active volcanoes on the island. As she walked, she found a rough, dark-colored rock. Later, Amanda found another rock. It was light-colored and had several layers. When she picked up this rock, some of the layers broke off.

> **Tip**
> What type of rock forms when volcanoes erupt?

A. What kind of rock was the dark-colored rock Amanda found? Explain your reasoning.

_____

_____

_____

_____

B. What kind of rock was the light-colored rock Amanda found? Explain your reasoning.

_____

_____

_____

_____

Chapter 6 Practice Set

Name _____

Date _____

Chapter 7 Practice Set

Use the illustration below to answer question 1.

1. The illustration shows a valley. Which of the following can carve a valley?

    Ⓐ earthquake

    Ⓑ glacier

    Ⓒ volcano

    Ⓓ wind

2. Which of these effects is most likely to result from an earthquake?

    Ⓐ uneven areas of ground

    Ⓑ lava

    Ⓒ gas clouds

    Ⓓ burned forests

Tip
Think of what happens during an earthquake. Remember that an earthquake is the shaking of Earth's surface.

Name _____ Date _____

Use the diagram below to answer question 3.

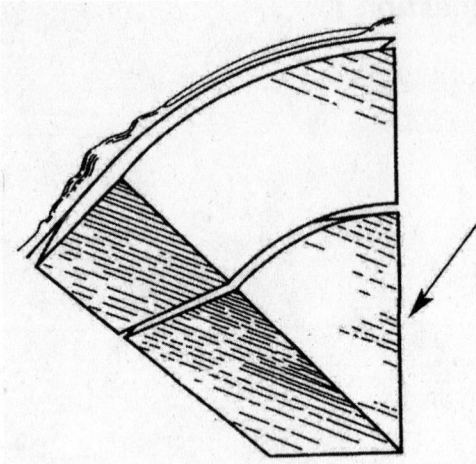

3. Earth has three layers. To which layer is the arrow pointing?

   Ⓐ crust

   Ⓑ core

   Ⓒ mantle

   Ⓓ outer

Tip
Remember the order of Earth's layers, from outside to inside.

Name _____ Date _____

Use the illustration below to answer question 4.

4. The illustration shows a plant growing in the cracks of a rock. What will eventually happen to the rock as the plant grows? Explain your reasoning.

_____

_____

_____

Tip
The illustration shows a type of weathering. How will weathering change the rock?

Name _____ Date _____

5. Jason visits his grandfather's farm each year. He has noticed that, over the years, the fence has become bent.

> **Tip**
> Think of processes that slowly change the land.

A. What could have caused the fence to bend? Explain your answer.

_____

_____

_____

B. Identify and describe the process that caused the fence to bend.

_____

_____

_____

Chapter 7 Practice Set

Name _____

Date _____

Chapter 8 Practice Set

1. Which of the following are pollutants?

    Ⓐ noise and conservation

    Ⓑ trees and chemicals

    Ⓒ chemicals and noise

    Ⓓ birds and noise

   > **Tip**
   > Look for the answer choice in which both items are pollutants.

Name _____ Date _____

Use the illustration below to answer question 2.

2. What type of resource is shown in the illustration?

   Ⓐ nonrenewable resource

   Ⓑ renewable resource

   Ⓒ reusable resource

   Ⓓ unlimited resource

   > **Tip**
   > Look at the picture and identify the resource. Is it a resource that can be reused or replaced?

3. Blake turns off the television when he is not watching it. He also turns off the lights when he leaves a room. Which resource is he conserving?

   Ⓐ electricity

   Ⓑ gasoline

   Ⓒ paper

   Ⓓ water

Name _____ Date _____

Use the illustration below to answer question 4.

4. What is the boy doing in the illustration? Explain your reasoning.

> **Tip**
> Recall the three things that people can do to use resources wisely. Which of these things is the boy doing?

_____

_____

_____

Chapter 8 Practice Set

Name _____ Date _____

5. Rocks, wood, and metals are building materials that come from the land.

   A. How do people use rocks, wood, and metals?

   > **Tip**
   > When you contrast two things, you tell how they are different.

   _____

   _____

   _____

   _____

   B. Contrast how land is used in a big city and in a small town.

   _____

   _____

   _____

   _____

   _____

   _____

   _____

Name _____

Date _____

Chapter 9 Practice Set

1. You wake up one morning and notice that the street is full of puddles. By the end of the day, the puddles are gone. What happened to the water during the day?

   Tip
   Think about the process by which liquid water changes to water vapor.

   Ⓐ The water boiled.

   Ⓑ The water condensed.

   Ⓒ The water evaporated.

   Ⓓ The water melted.

2. What happens during the condensation stage of the water cycle?

   Ⓐ Liquid water turns to water vapor.

   Ⓑ Clouds form from water vapor.

   Ⓒ Liquid water falls as rain.

   Ⓓ Liquid water falls as sleet, snow, and hail.

Name _____ Date _____

Use the illustration below to answer question 3.

3. The illustration shows water in one of its forms. What can you tell about the temperature of the water by looking at the picture?

Tip
Think about the form of water shown in the picture.

Ⓐ The temperature is below 0°C.

Ⓑ The temperature is above 0°C.

Ⓒ The temperature is about 32°C.

Ⓓ The temperature is exactly 100°C.

Name _____ Date _____

4. Alicia plants a small plant in a plastic container. She waters the plant and then covers the container with clear plastic wrap. She places the plant on a sunny windowsill. The next morning, she notices drops of water on the inside of the plastic wrap. Why does water appear on the plastic wrap?

> **Tip**
> What causes water droplets to form on a surface?

_____

_____

_____

Name _____ Date _____

Use the illustration below to answer question 5.

5. Will and his mother made the model of the water cycle shown in the illustration.

   Tip
   Match each part of Will's model with a part of the water cycle.

   A. How does Will's model show evaporation and condensation?

   _____
   _____
   _____
   _____

   B. How can Will use his model to show precipitation? Explain.

   _____
   _____
   _____
   _____

Name _____

Date _____

Chapter 10 Practice Set

1. How long does it take Earth to rotate one full turn on its axis?

    Ⓐ 1 day

    Ⓑ 7 days

    Ⓒ 29 days

    Ⓓ 365 days

Use the diagram below to answer question 2.

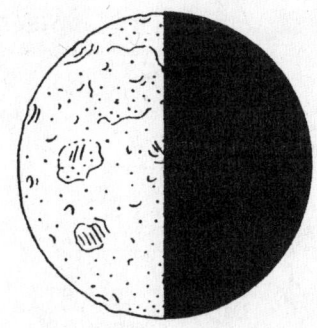

2. The diagram shows the moon in the third-quarter phase. About how long will it take the moon to enter this phase again?

    Ⓐ 24 hours

    Ⓑ 7 days

    Ⓒ 29 days

    Ⓓ 1 year

> **Tip**
> Think of the time between one third-quarter moon and the next third-quarter moon, or a whole cycle.

Name _____ Date _____

3. Constellations are groups of stars that appear to form shapes in the night sky. What do constellations suggest about the sun and other stars?

> **Tip**
> Think about how small the stars look compared to the sun.

Ⓐ The sun is a planet instead of a star.

Ⓑ All stars are in our solar system.

Ⓒ There are many other stars, but they are far away.

Ⓓ All other stars are brighter than the sun.

Name _____ Date _____

Use the diagram below to answer question 4.

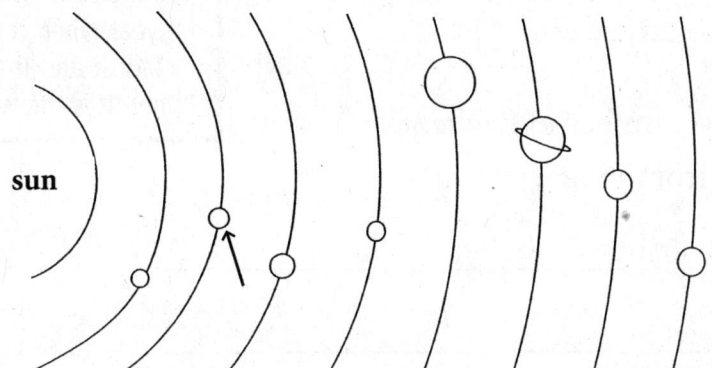

4. The diagram shows our solar system. To which planet is the arrow pointing? Explain how you know.

Tip
Write the names of the planets in order, and then identify the second planet away from the sun.

_____
_____
_____

Name _____

Chapter 10 Practice Set

Name _____ Date _____

5. Earth revolves around the sun and rotates on its axis. This causes changes that we can see and feel.

Tip
Think of the times of the year when it is the coldest and the times when it is the warmest.

A. What happens to the number of daylight hours as the year goes from January to June?

_____

_____

_____

B. What happens to temperatures as we move through the seasons of spring, summer, fall, and winter?

_____

_____

_____

_____

Name _____

Date _____

Chapter 11 Practice Set

1. Tim put four objects in a tank of water. All but one sank to the bottom of the tank. Which object most likely floated?

    Ⓐ rock

    Ⓑ beach ball

    Ⓒ golf ball

    Ⓓ marble

Tip
Think of the density of each object. Matter that is less dense than water will float.

2. What are the three states of matter?

    Ⓐ solid, liquid, and water

    Ⓑ liquid, air, and gas

    Ⓒ solid, liquid, and gas

    Ⓓ mass, density, and volume

3. In which example would the water molecules be moving most quickly?

> **Tip**
> Find the picture in which water receives the greatest amount of heat.

A

B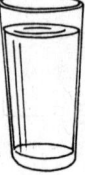

C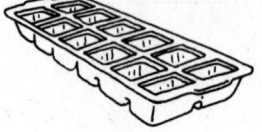

D

Name _____ Date _____

Use the diagram below to answer question 4.

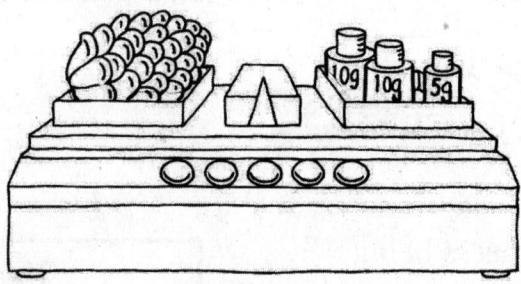

4. The diagram shows a pinecone on a balance. What is the mass of the pinecone? Explain how you found the mass.

> **Tip**
> The pans are level, so the masses on each side are equal.

_____
_____
_____
_____

Chapter 11 Practice Set 47

Name _____ Date _____

Use the illustration below to answer question 5.

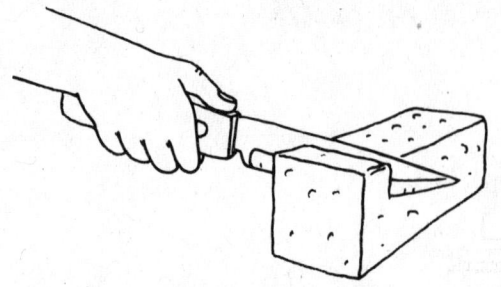

5. The illustration shows a block of cheese being changed.

    A. Is this a physical change or a chemical change? Explain your answer.

    Tip
    Remember the difference between a physical change and a chemical change.

    _____
    _____
    _____
    _____

    B. If you compared the mass of the cheese before it was sliced to the mass of the slices, what would you find? Explain how you know.

    _____
    _____
    _____
    _____

48 Chapter 11 Practice Set

Name _____

Date _____

Chapter 12 Practice Set

Use the illustration below to answer question 1.

1. The illustration shows a ball rolling down a hill. What kind of energy does the ball have?

    Ⓐ potential energy

    Ⓑ electric energy

    Ⓒ kinetic energy

    Ⓓ magnetic energy

2. Mrs. Jones is hoping that the flowers in her garden will bloom. The plants need energy to grow. Where will the plants get the energy they need?

    Ⓐ the air

    Ⓑ the sun

    Ⓒ the moon

    Ⓓ the water

Tip
What do plants need to grow? Which choice is an energy source?

(page 1 of 4)

Chapter 12 Practice Set 49

Name _____ Date _____

3. Which of the following is a nonrenewable resource?

> **Tip**
> Recall the meaning of *nonrenewable*.

Ⓐ electricity

Ⓑ heat

Ⓒ wind

Ⓓ oil

Name _____ Date _____

Use the illustration below to answer question 4.

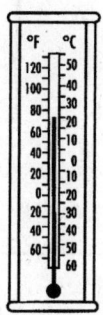

4. The illustration shows a thermometer. What type of energy does this tool measure? What units are used to measure this type of energy?

Tip
Temperature is a measure of how hot or cold something is.

_____

_____

_____

Name _____ Date _____

5. Sam has just put new batteries in his remote-control car. He has not had time to play with the car yet, so it is just sitting on a shelf in his room.

> **Tip**
> Think of the position and motion of the car. How do batteries cause the car to move?

A. What type of energy do the car and the batteries in the car have? Explain how you know.

_____

_____

_____

_____

B. If Sam wanted to change the car's energy, what could he do?

_____

_____

_____

_____

Chapter 12 Practice Set

Name _____

Date _____

Chapter 13 Practice Set

Use the illustration below to answer question 1.

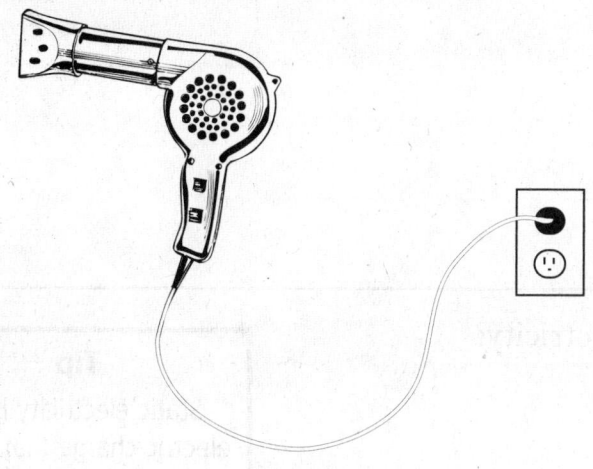

1. What type of electricity runs this object?

   Ⓐ static electricity

   Ⓑ current electricity

   Ⓒ magnetic electricity

   Ⓓ solar electricity

Tip
The electricity is coming through the wire from the outlet.

(page 1 of 4) Chapter 13 Practice Set **53**

Name _____ Date _____

2. Which of the following will a magnet attract?

　Ⓐ plastic paper clip

　Ⓑ paper cup

　Ⓒ steel paper clip

　Ⓓ rubber ball

3. Which is an example of static electricity?

　Ⓐ lightning

　Ⓑ a lamp

　Ⓒ a magnet

　Ⓓ rain

Tip
Static electricity is an electric charge that builds up in an object.

Name _____ Date _____

Use the illustration below to answer question 4.

 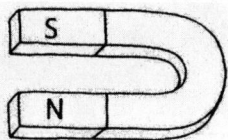

4. What will these two magnets do when they are brought closer together? Explain your answer.

Tip
Look at how the poles on the two magnets are aligned.

_____
_____
_____
_____

Chapter 13 Practice Set

Name _____ Date _____

5. Michael has a steel screw, wire, and a battery.

   A. How can he use those items to make an electromagnet?

   > **Tip**
   > Think about what you would need to do to make the screw a magnet.

   _____
   _____
   _____
   _____
   _____

   B. How is an electromagnet different from a regular magnet?

   _____
   _____
   _____
   _____

Name _____

Date _____

Chapter 14 Practice Set

1. Which is an example of a good insulator?

   Ⓐ iron

   Ⓑ cloth

   Ⓒ aluminum

   Ⓓ copper

   Tip
   An insulator is a material that is a poor conductor of heat.

Use the graph below to answer question 2.

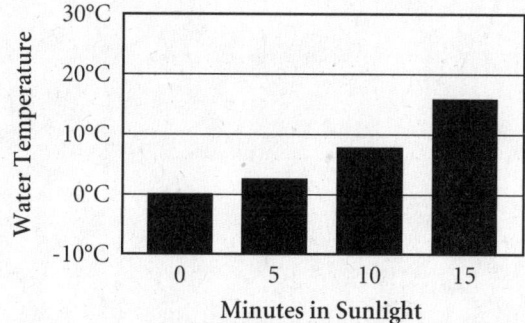

2. In a class experiment, a cup of ice was placed in sunlight. Students took the temperature of the water every five minutes. The graph shows the results. Which statement best explains the graph?

   Ⓐ The water cooled the air around it.

   Ⓑ Water temperature is not affected by sunlight.

   Ⓒ Heat from the sun warmed the water.

   Ⓓ The longer the water was in the sunlight, the cooler it got.

3. Which object is opaque?

> **Tip**
> Objects that are opaque do not let light pass through them.

Ⓐ

Ⓑ

Ⓒ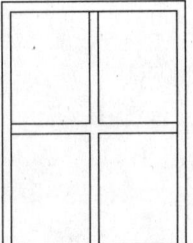

Ⓓ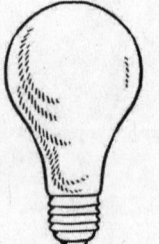

Name _____ Date _____

4. What two naturally occurring things are needed to see a rainbow outside? Explain your answer.

> **Tip**
> Think about what a prism does to light.

_____

_____

_____

Name _____ Date _____

5. Holly wants to make white light.

A. What materials will she need to make white light?

> **Tip**
> What three colors make white light? Think about the types of objects you can use to make the colored lights.

_____

_____

_____

_____

B. Explain how to use the materials to make the light.

_____

_____

_____

_____

Name _____
Date _____

Chapter 15 Practice Set

1. Which of the following units measure only distance?

    Ⓐ centimeters and meters

    Ⓑ inches and minutes

    Ⓒ minutes and hours

    Ⓓ miles and seconds

Tip
Look for the answer choice in which both are units of distance.

Use the illustration below to answer question 2.

2. The illustration shows a girl hitting a baseball with a bat. The bat changes the speed and the direction of the ball. What can you conclude about the relationship between the bat and ball?

    Ⓐ The bat applies friction to the ball.

    Ⓑ The bat applies gravity to the ball.

    Ⓒ The bat applies a force to the ball.

    Ⓓ The bat applies electricity to the ball.

Name _____ Date _____

Use the diagram below to answer question 3.

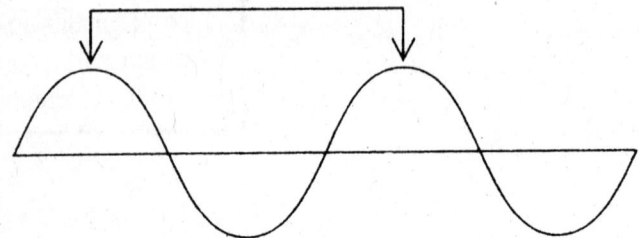

3. Lyle is using the diagram to measure the length of waves. The arrows show the two points he is measuring. What are these points called?

Tip
What are the parts of a wave?

Ⓐ troughs

Ⓑ valleys

Ⓒ crests

Ⓓ wavelengths

Name _____ Date _____

4. When Brian opened his closet door, a backpack fell from the top shelf. How do you know a force has acted on the backpack?

> **Tip**
> What causes an object to move?

_____

_____

_____

_____

Name _____ Date _____

Use the illustrations below to answer question 5.

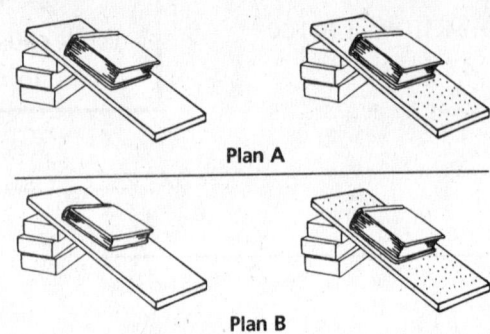

5. Paula is investigating the force of friction. She set up two ramps: one with a smooth surface and one with a sandpaper surface. She has two plans, as shown.

Tip
Which plan allows a fair comparison?

A. Which plan is better for investigating the force of friction? Explain your reasoning.

_____

_____

_____

_____

_____

B. When Paula places the book on the ramp with a sandpaper surface, the book does not move. Why does this happen?

_____

_____

_____

_____

Chapter 15 Practice Set

Name _____

Date _____

Chapter 16 Practice Set

1. Cass tells her mother that scientists can measure work. Which of these must be known to measure work?

    > **Tip**
    > Think about what *work* means in science.

    Ⓐ how much something weighs

    Ⓑ how far something moves

    Ⓒ how tall something is

    Ⓓ how hard something is to figure out

Use the illustration to answer question 2.

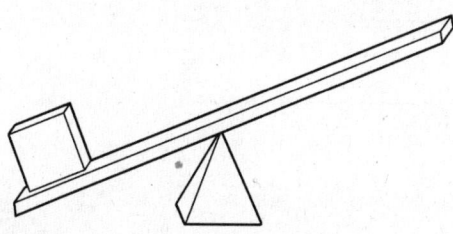

2. What kind of simple machine is shown in the illustration?

    Ⓐ wheel-and-axle

    Ⓑ inclined plane

    Ⓒ pulley

    Ⓓ lever

(page 1 of 4)

Name _____ Date _____

3. Max uses a wheelchair. Which simple machine does Max most likely use to get past three steps at the front entrance of his school?

Tip
Picture Max's wheelchair in your mind. What would he need to get from one level to the next?

Ⓐ inclined plane

Ⓑ lever

Ⓒ screw

Ⓓ wedge

Name _____ Date _____

Use the table to answer question 4.

Simple Machine	Example
lever	
wheel-and-axle	
pulley	
wedge	
screw	
inclined plane	

4. Carlos is studying simple machines. He is making a chart to show which ones he finds in his home. Write an example Carlos could use, next to the name of each simple machine.

Tip
Think of how a simple machine moves and how it can help people.

_____

_____

_____

_____

Chapter 16 Practice Set 67

Name _____ Date _____

Use the illustrations to answer question 5.

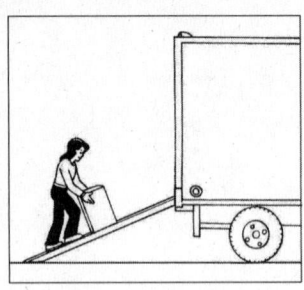 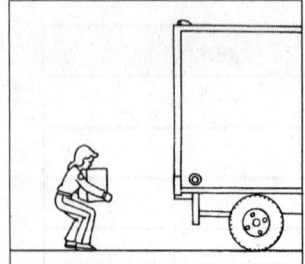

        A   **B**

5. Peter was asked to draw two pictures that involved work. He drew these pictures of two women moving boxes. The boxes are identical in weight, and the trucks are identical.

> **Tip**
> Identify the woman who uses less force over a greater distance to move the box.

A. Do you think the woman in Picture A is doing more work? Explain your answer.

_____

_____

_____

_____

B. Which woman has an easier job? Explain your answer.

_____

_____

_____

_____

_____

_____

Chapter 16 Practice Set

1. Which of these is a way you use minerals in daily life?

   Ⓐ drinking milk

   Ⓑ putting blueberries in muffins

   Ⓒ putting salt on food

   Ⓓ drinking tea

S4.D.1.2

Use the diagram below to answer question 2.

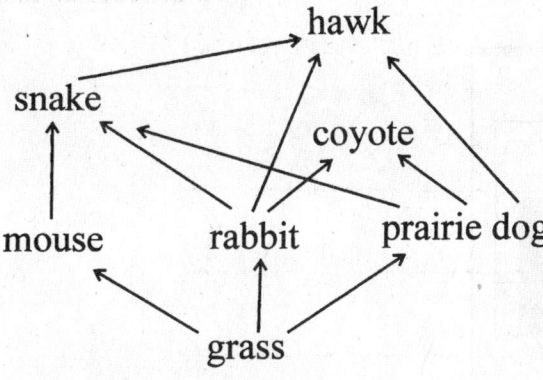

2. If harmful chemicals were sprayed on the grass, which organisms in the food web would be affected first?

   Ⓐ snake, hawk, coyote

   Ⓑ coyote, mouse, snake

   Ⓒ prairie dog, hawk, rabbit

   Ⓓ mouse, rabbit, prairie dog

S4.B.3.2

Name _____

Date _____

Practice Test 1

3. Think about the difference between kinetic and potential energy. Which of these objects has kinetic energy?

    Ⓐ a pebble rolling down a hill

    Ⓑ the food on a dinner plate

    Ⓒ a soccer ball resting on the ground

    Ⓓ the gum balls in a gum machine

S4.C.2.1

Use the bar graph below to answer question 4.

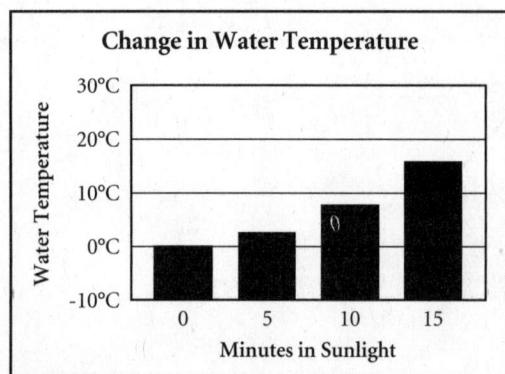

4. Sam placed a cup of ice in the sunlight. He took the temperature of the water every five minutes. The graph shows his results. What can you infer about water temperature and sunlight?

    Ⓐ The water temperature is not affected by sunlight.

    Ⓑ The longer the water is in the sun, the cooler it gets.

    Ⓒ The longer the water is in the sun, the warmer it gets.

    Ⓓ The water temperature stayed the same.

S4.A.2.1

Name _____

Date _____

Practice Test 1

Use the illustrations below to answer question 5.

soil   metal rod

5. Which term describes all of these resources?

    Ⓐ reusable resources

    Ⓑ unlimited resources

    Ⓒ renewable resources

    Ⓓ nonrenewable resources

S4.D.1.2

6. Which statement about animals is considered a fact?

    Ⓐ Animals need energy to survive.

    Ⓑ Animals that lay eggs are interesting.

    Ⓒ Animals that live in the ocean are very intelligent.

    Ⓓ Animals with fur should not be hunted.

S4.A.1.1

7. Which of these is a nonliving feature in a forest ecosystem?

Ⓐ

Ⓑ

Ⓒ

Ⓓ

S4.A.3.1

8. Which statement describes a pattern of growth observed in deciduous plants?

Ⓐ Leaves are usually lost in the fall.

Ⓑ Seeds are produced in cones.

Ⓒ Flowers produce spores.

Ⓓ Leaves are needlelike.

Name _____

Date _____

Practice Test 1

9. Which type of clouds are the highest clouds and look like feathers?

   Ⓐ cumulus

   Ⓑ cirrus

   Ⓒ stratus

   Ⓓ rain

S4.D.2.1

Use the diagram below to answer question 10.

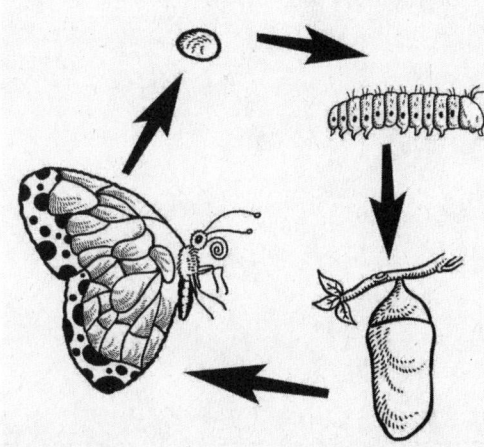

10. Which of the following is modeled by the diagram?

    Ⓐ a food web

    Ⓑ a food chain

    Ⓒ a size comparison

    Ⓓ a life cycle

S4.D.2.1

(page 5 of 18)   Practice Test 1   73

11. Which is an example of gravity?

    Ⓐ pulling open a drawer

    Ⓑ rubbing hands together

    Ⓒ a ball falling to the ground

    Ⓓ pushing a shopping cart

S4.C.3.1

Use the illustration below to answer question 12.

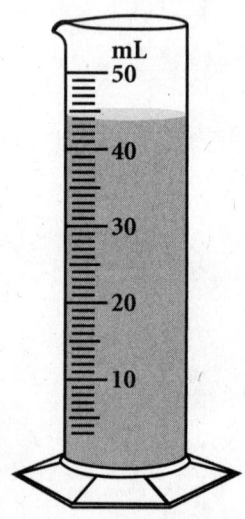

12. What is the volume of liquid in this graduated cylinder?

    Ⓐ 35 mL

    Ⓑ 40 mL

    Ⓒ 45 mL

    Ⓓ 50 mL

S4.A.2.2

13. Which picture shows someone changing the motion of an object with a pushing force?

S4.C.3.1

14. How do beavers change the environment to build a dam?

Ⓐ They eat small plants.

Ⓑ They cut down trees.

Ⓒ They dig underground burrows.

Ⓓ They pollute the water.

S4.B.3.2

Use the illustration below to answer question 15.

15. The illustration shows precipitation falling from a cloud. What are both the cloud and precipitation made of?

   Ⓐ liquid water

   Ⓑ ice

   Ⓒ water vapor

   Ⓓ air

16. Which human activity causes air pollution?

   Ⓐ burning fossil fuels

   Ⓑ oil spilled from tanker ships

   Ⓒ using chemical fertilizers

   Ⓓ not using a trash can

Name _____

Date _____

Practice Test 1

Use the illustration below to answer question 17.

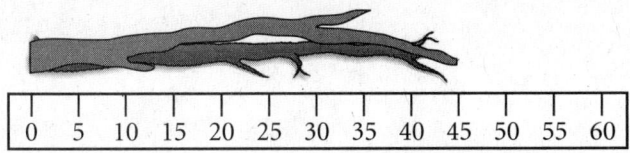

17. What is the length of this branch?

    Ⓐ 30 cm

    Ⓑ 35 cm

    Ⓒ 45 cm

    Ⓓ 53 cm

S4.A.2.2

18. Which term describes an adaptation some animals use to survive the cold months of winter?

    Ⓐ mimicry

    Ⓑ hibernation

    Ⓒ reproduction

    Ⓓ metamorphosis

S4.B.2.1

Name _____

Date _____

Practice Test 1

19. Porcupines have long quills. What function do the quills provide for a porcupine?

   Ⓐ attract a mate

   Ⓑ store water

   Ⓒ attract predators

   Ⓓ protect them

S4.B.2.1

Use the illustrations below to answer question 20.

20. The illustrations show a fern plant and a daisy plant. How are these plants different?

   Ⓐ The fern has leaves. The daisies do not.

   Ⓑ The fern has a stem. The daisies do not.

   Ⓒ The daisies have flowers. The fern does not.

   Ⓓ The daisies have roots. The fern does not.

S4.B.1.1

Use the illustration below to answer question 21.

21. Which of the following forms a valley?

    Ⓐ swirling tornado

    Ⓑ moving glacier

    Ⓒ erupting volcano

    Ⓓ eroding wind

S4.D.1.1

22. Sometimes trash is dumped into the ocean. How does recycling trash help animals that live in the ocean?

    Ⓐ By reducing water pollution

    Ⓑ By providing shelter for ocean animals

    Ⓒ By increasing the temperature of the ocean

    Ⓓ By providing more food

S4.A.1.3

23. Which sandwich will stay the freshest and be the safest to eat?

S4.A.2.1

24. It takes the moon about 29 days to orbit Earth. How long does it take for Earth to orbit the sun?

- Ⓐ one hour
- Ⓑ one week
- Ⓒ one month
- Ⓓ one year

S4.D.3.1

Name _____

Date _____

Practice Test 1

Use the illustration below to answer question 25.

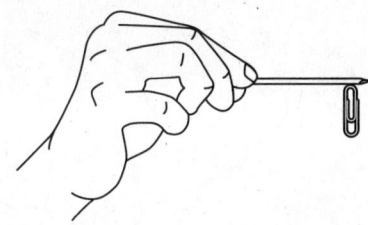

25. What can you infer from this illustration?

   Ⓐ The paper clip is plastic.

   Ⓑ The paper clip will soon fall.

   Ⓒ The nail has magnetic poles.

   Ⓓ The nail repels the paper clip.

S4.C.3.1

Use the graph below to answer question 26.

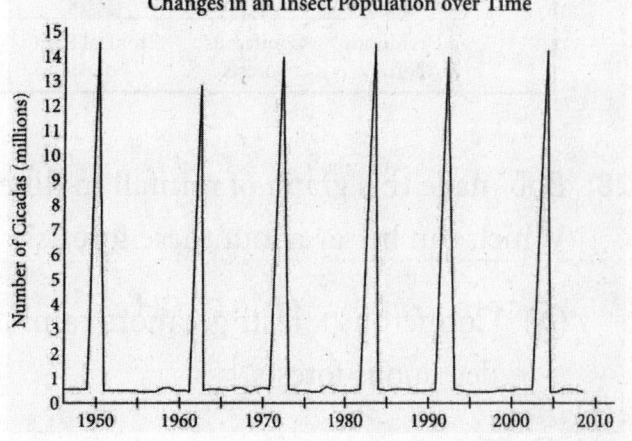

26. This graph shows when a certain type of insect emerges from dormancy to mate.

   What is the estimated year the insects will next appear?

   Ⓐ 2000

   Ⓑ 2005

   Ⓒ 2015

   Ⓓ 2030

S4.A.3.3

27. Which is the advantage of a giraffe's long neck?

   Ⓐ to help the giraffe sleep

   Ⓑ to help the giraffe reach its food

   Ⓒ to help the giraffe swim

   Ⓓ to help the giraffe get more sun

S4.B.2.1

Use the bar graph below to answer question 28.

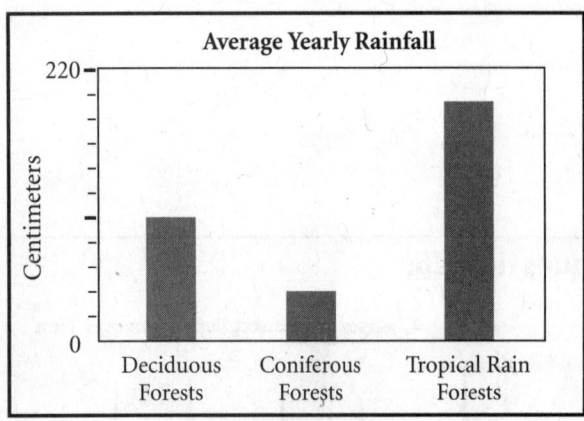

28. Bob made this graph of rainfall in different kinds of forests. Which can he say about these forests?

   Ⓐ Coniferous forests get more rainfall than deciduous forests.

   Ⓑ No animals can survive in these forests.

   Ⓒ Rain forests get more rainfall than deciduous forests.

   Ⓓ Plants do not compete for resources in these forests.

S4.B.3.1

Name _____

Date _____

Practice Test 1

Use the illustrations below to answer question 29.

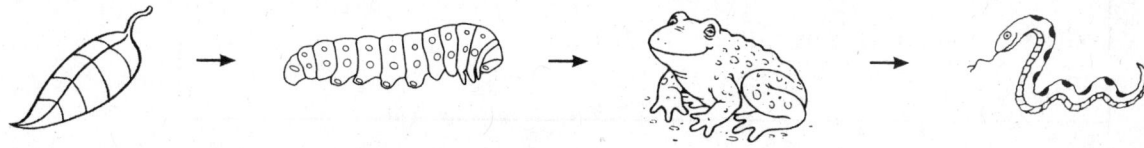

29. Study the food chain shown in the diagram.

A. What is the producer in this food chain? How does the producer get food?

B. What would happen if the frog were removed from this food chain?

S4.A.1.3

Name _____

Date _____

Practice Test 1

30. Ducks live in areas with a lot of water.

   A. What are two inherited characteristics that help ducks survive in a water environment?

   B. What are two learned characteristics that help ducks survive in a water environment?

Name _____

Date _____

Practice Test 1

31. Matter can be classified according to its state.

A. What are the three states of matter?

B. Give an example of matter in each state.

S4.C.1.1

Name _____

Date _____

Practice Test 1

Use the illustration below to answer question 32.

32. The illustration shows a car, a modern tool that many people depend on.

A. Identify the kind of resources used to make the car.

A. How have cars affected your daily life? Identify one positive effect and one negative effect.

Name _____

Date _____

Practice Test 2

1. Soil is a mixture of small pieces of rock and decayed plant and animal matter. Which of these can help break large rocks into the tiny bits found in soil?

    Ⓐ water only

    Ⓑ wind only

    Ⓒ plants, air, and animals

    Ⓓ wind, water, and plants

S4.D.1.1

Use the illustration below to answer question 2.

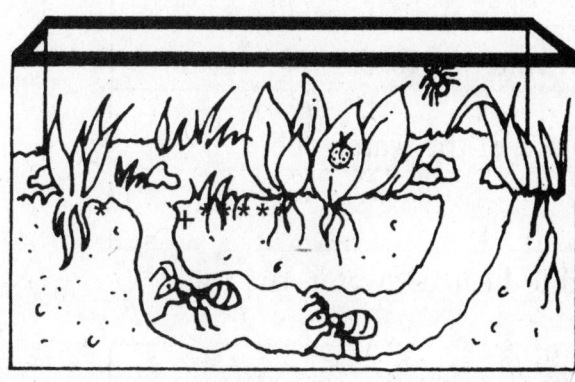

2. What is a nonliving part of the ant farm shown in the illustration?

    Ⓐ the plants

    Ⓑ the soil

    Ⓒ the ants

    Ⓓ the bacteria in the soil

S4.B.3.1

Name _____

Date _____

Practice Test 2

Use the illustration below to answer question 3.

3. Which hypothesis could be tested with the set-up shown above?

Ⓐ The water in the pot is at 45°C.

Ⓑ Water at 45° C is hotter than water at 40°C.

Ⓒ If the thermometer is taken out of the water, the temperature will fall.

Ⓓ When a different pot is used to heat the water, the water will get hotter.

S4.A.2.2

4. What is an effect of the tilt of Earth on its axis?

Ⓐ The size of Earth changes during the year.

Ⓑ Earth rotates faster and then slower during the year.

Ⓒ The length of a year changes from year to year.

Ⓓ Seasons change in different parts of Earth during the year.

S4.D.3.1

Use the illustration below to answer question 5.

5. What type of energy causes the butter on the pancakes to melt?

   Ⓐ electrical

   Ⓑ heat

   Ⓒ magnetism

   Ⓓ potential

S4.C.2.1

6. Whales live in the ocean. They give birth to live young and feed milk to their babies. What type of vertebrate is a whale?

   Ⓐ mammal

   Ⓑ fish

   Ⓒ amphibian

   Ⓓ reptile

S4.B.1.1

Use the illustration below to answer question 7.

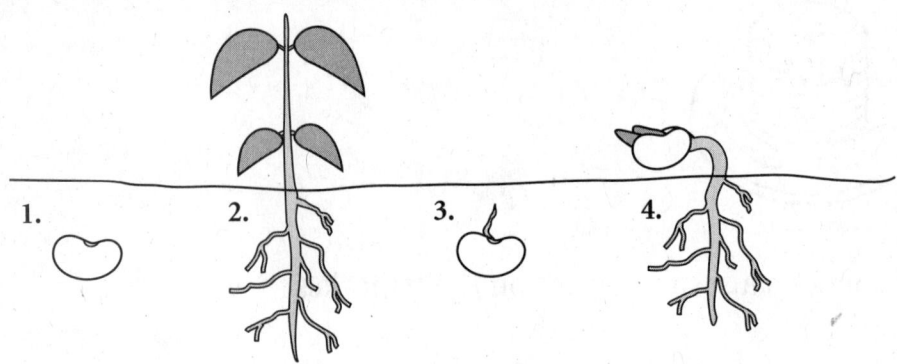

7. The illustrations show four different stages in the life of a bean plant. Which is the correct order of the stages, from earliest to latest?

   Ⓐ 2, 4, 3, 1

   Ⓑ 1, 3, 4, 2

   Ⓒ 4, 2, 3, 1

   Ⓓ 3, 1, 2, 4

S4.A.3.3

8. A lizard's skin feels like leaves, and is the same color. How does its skin help the lizard survive?

   Ⓐ It makes it harder for predators to find the lizard.

   Ⓑ It makes it harder for the lizard to find food.

   Ⓒ It makes it easier for the lizard to stay warm.

   Ⓓ It makes it possible for the lizard to make its own food.

S4.B.2.1

Use the illustrations below to answer question 9.

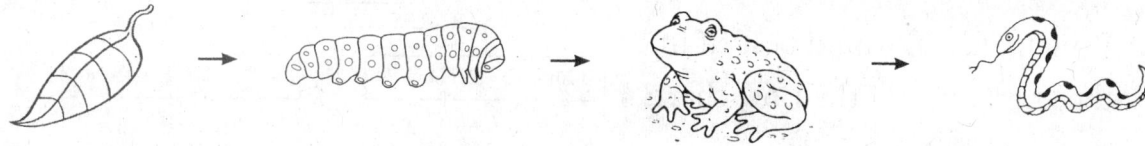

9. The pictures show four living organisms. Which is the correct order of these organisms in a food chain?

   Ⓐ snake, frog, plant, caterpillar

   Ⓑ caterpillar, snake, frog, plant

   Ⓒ plant, snake, frog, caterpillar

   Ⓓ plant, caterpillar, frog, snake

10. What is the name for a change in the position of an object?

    Ⓐ motion

    Ⓑ force

    Ⓒ magnetism

    Ⓓ friction

Name _____

Date _____

Practice Test 2

Use the flow chart below to answer question 11.

11. Which animal would best fit in the empty box of the flow chart?

    Ⓐ

    Ⓑ

    Ⓒ

    Ⓓ

S4.B.1.1

12. About 2,200 years ago, the Greek mathematician Archimedes invented a machine called the Archimedes' screw. A handle turned the screw and raised water from underground wells. How might the Archimedes' screw have changed the way people lived?

    Ⓐ People could build taller buildings.

    Ⓑ People could dig deeper holes.

    Ⓒ People could grow more food.

    Ⓓ People could cut down bigger trees.

S4.A.1.1

Name _____

Date _____

Practice Test 2

Use the map below to answer question 13.

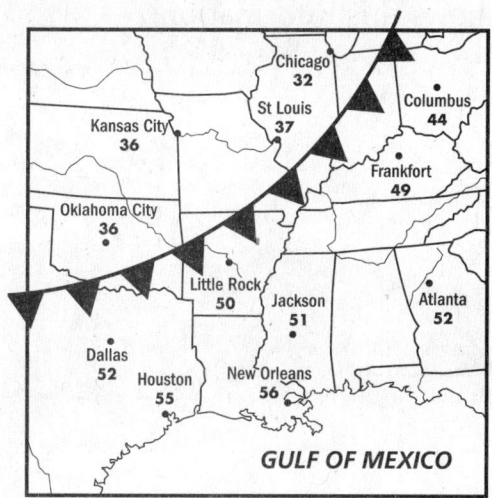

13. What does the weather symbol on the map indicate?

   Ⓐ high

   Ⓑ low

   Ⓒ cold front

   Ⓓ warm front

S4.D.2.1

14. What motion does Earth complete in 24 hours?

   Ⓐ one revolution

   Ⓑ one orbit

   Ⓒ one rotation

   Ⓓ one lunar cycle

S4.D.3.1

15. On Tuesday, the amount of rain that fell increased each hour during the morning. Which graph shows this information?

A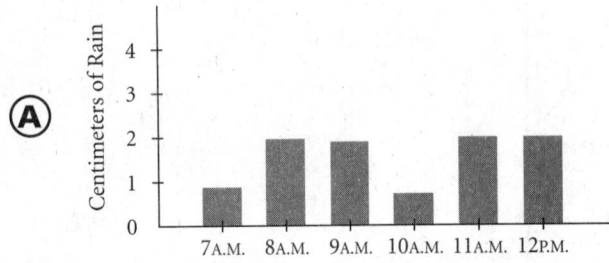

B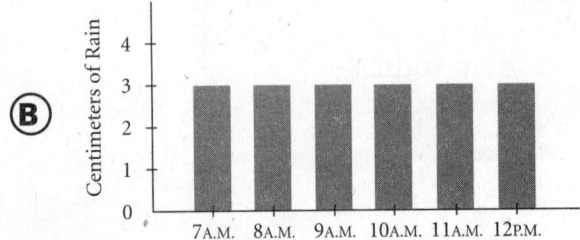

C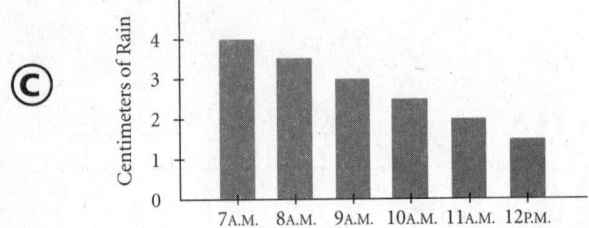

D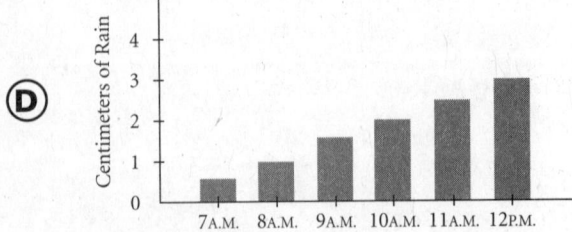

Name _____

Date _____

Practice Test 2

16. Which of the following do all humans need to live?

（A）milk

（B）water

（C）meat

（D）vegetables

S4.B.1.1

17. Suppose a large number of rainforest plants were cut down. What impact would this have on the environment?

（A）There would be less light on the rainforest floor.

（B）There would be less water in the ground.

（C）The soil would have fewer nutrients.

（D）Less oxygen would be released into the air.

S4.A.1.3

18. What is one way a blizzard changes the habitats of plants and animals living in a forest ecosystem?

（A）Snow covers plants that animals need for food.

（B）Moisture from the snow makes the trees grow taller.

（C）Colder temperatures make it harder for plants to make food.

（D）The white snow makes it harder for animals to see.

S4.B.3.2

Name _____

Date _____

Practice Test 2

Use the diagrams below to answer question 19.

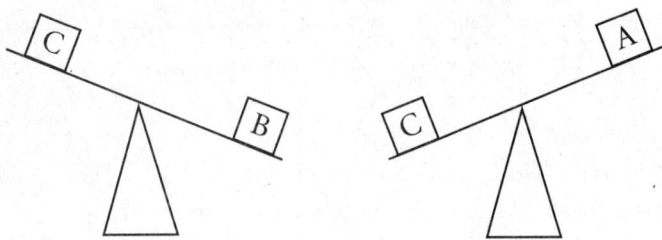

19. Which box has the greatest mass?

    Ⓐ box A

    Ⓑ box B

    Ⓒ box C

    Ⓓ all are equal

S4.C.1.1

20. Your rabbit has babies. Some of the baby rabbits have white fur. Others have black fur. Some even have black-and-white fur. Which word explains the different colors?

    Ⓐ growing

    Ⓑ breathing

    Ⓒ reproduction

    Ⓓ variation

S4.B.2.2

21. Which of these is an SI unit used to measure temperature?

Ⓐ degrees Fahrenheit

Ⓑ degrees Celsius

Ⓒ kilometers

Ⓓ gallons

S4.A.2.2

Use the illustration below to answer question 22.

22. You look at the night sky and see a full moon. Which phase of the moon will you see in about 29 days?

Ⓐ new moon

Ⓑ half moon

Ⓒ full moon

Ⓓ quarter moon

S4.D.3.1

23. Which object will float in a tub of water?

    Ⓐ a bowling ball

    Ⓑ a rock

    Ⓒ a feather

    Ⓓ a book

S4.C.1.1

Use the diagram below to answer question 24.

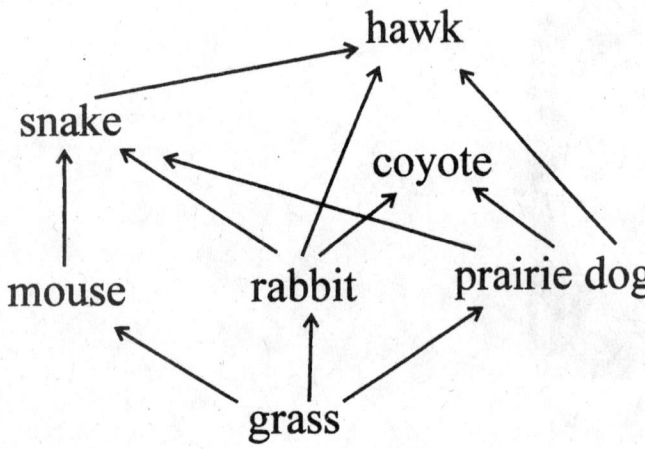

24. What do the arrows in the diagram show?

    Ⓐ transfer of energy

    Ⓑ animals' homes

    Ⓒ sizes of animals

    Ⓓ animals' life spans

25. Which set of magnetic poles will attract one another?

Ⓐ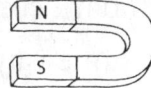

Ⓑ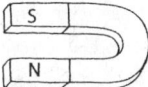

Ⓒ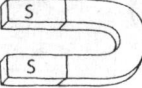

Ⓓ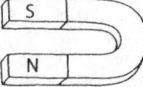

S4.C.3.1

Use the bar graph below to answer question 26.

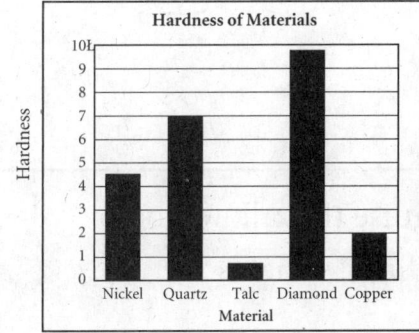

26. The graph compares the hardness of several minerals. Which statement is a fact about the information in the graph?

Ⓐ Copper should not be worn as jewelry.

Ⓑ Talc is softer than quartz.

Ⓒ Diamonds are everybody's favorite mineral.

Ⓓ Quartz is collected as frequently as copper.

S4.A.1.1

27. Alma started an experiment with two identical plants. She put Plant 1 in a sunny window and Plant 2 in a dark corner of the same room. She watered the plants the same way. At the end of 8 weeks, she observed the plants. Which picture shows how the plants probably changed?

    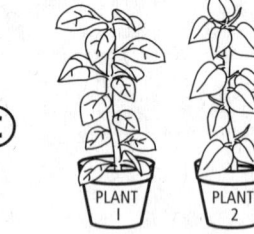

 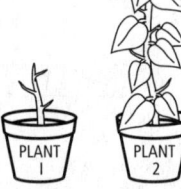   

S4.A.2.1

28. One type of adaptation is a natural disguise that allows some animals to blend in with their surroundings. What is the name of this adaptation?

Ⓐ migration

Ⓑ mimicry

Ⓒ hibernation

Ⓓ camouflage

S4.B.2.1

Name _____

Date _____

Practice Test 2

29. A small stream flows by your school. One year, sand carried by the stream blocked the flow of water at a place near the school. As a result, the school was flooded.

   A. Describe the parts of the stream's system.

   B. Explain how these parts interacted to cause a flood.

Name _____

Date _____

Practice Test 2

30. Mary built a container to collect "gray water" from her house. Gray water is waste water from showers, clothes washers, and sinks. It can be used to wash cars or water plants.

   A. Describe a simple experiment for Mary to test whether gray water is safe for her houseplants.

   B. Explain how using gray water can benefit the environment.

Name _____

Date _____

Practice Test 2

Use the diagram below to answer question 31.

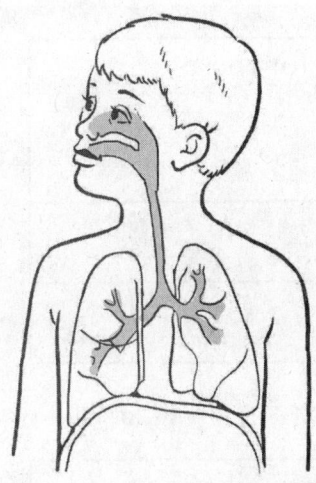

31. Study the diagram of the human body system.

   A. What is the name of the system shown in the diagram? What are two parts of the system?

   _____
   _____
   _____

   B. What is the function of this system?

   _____
   _____
   _____

S4.A.3.1

Name _____

Date _____

Practice Test 2

32. Marcus brushes his teeth twice a day and takes a shower every morning.

   A. Identify two resources Marcus relies on for these activities.

   B. What are two ways Marcus could conserve his use of these resources while performing the activities?

Use the illustrations below to answer question 1.

1. What is the difference in temperature between the two containers?

    Ⓐ 45°C

    Ⓑ 50°C

    Ⓒ 75°C

    Ⓓ 100°C

S4.A.2.2

2. What is the difference between an invertebrate and a vertebrate?

    Ⓐ Vertebrates have legs.

    Ⓑ Invertebrates have eyes.

    Ⓒ Vertebrates have a backbone.

    Ⓓ Invertebrates are mammals.

S4.B.1.1

Use the illustration below to answer question 3.

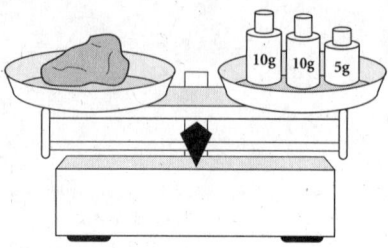

3. What is the mass of the rock?

   Ⓐ 5 g

   Ⓑ 10 g

   Ⓒ 15 g

   Ⓓ 25 g

S4.C.1.1

Use the pictures below to answer question 4.

4. In which ecosystem are you most likely to find these animals?

   Ⓐ tundra

   Ⓑ pond

   Ⓒ ocean

   Ⓓ desert

S4.B.2.1

Name _____

Date _____

Practice Test 3

5. What causes day and night on Earth?

    Ⓐ the tilt of Earth's axis

    Ⓑ Earth's rotation

    Ⓒ Earth's revolution around the sun

    Ⓓ the angle of the sun

S4.D.3.1

6. Which of the following is a type of technology used by farmers?

    Ⓐ nutrient-rich soil

    Ⓑ a tractor

    Ⓒ water

    Ⓓ seeds

S4.A.1.1

7. Marcus wants to make a model of an ecosystem. He puts some moist soil in the bottom of a large glass jar, then he adds some plants and a small dish of water. Which type of ecosystem has Marcus modeled?

    Ⓐ desert

    Ⓑ ocean

    Ⓒ tundra

    Ⓓ pond

S4.A.3.2

8. Your class has set up a weather station. On Tuesday, you checked the rain gauge. The number of centimeters of rain decreased each hour. Which graph shows this information?

Ⓐ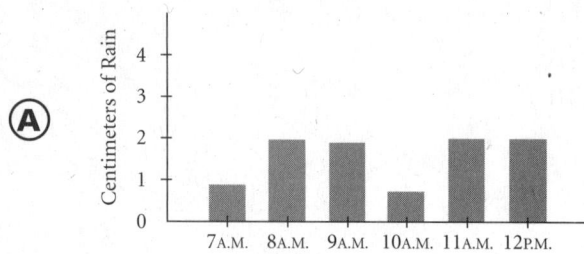

Ⓑ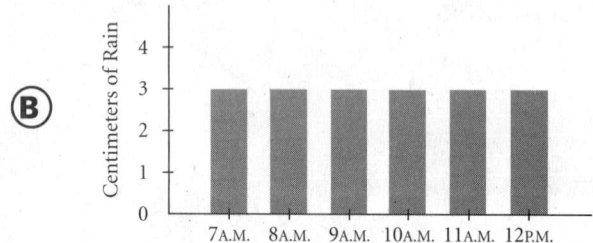

Ⓒ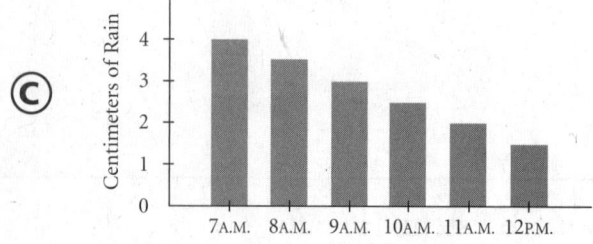

Ⓓ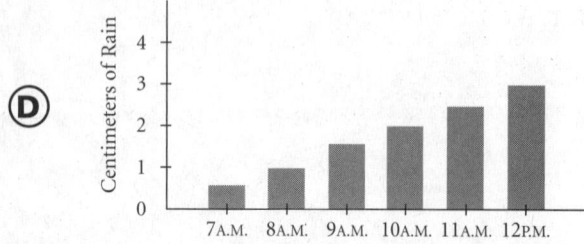

Name _____

Date _____

Practice Test 3

9. On a hot day, you look up and see a dark cloud that is rotating. What kind of weather may be about to happen?

    Ⓐ hurricane

    Ⓑ snowstorm

    Ⓒ tornado

    Ⓓ sunny day

S4.D.2.1

10. Which tool would a scientist use to study the parts of a flower?

    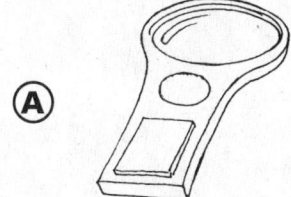
    Ⓐ

    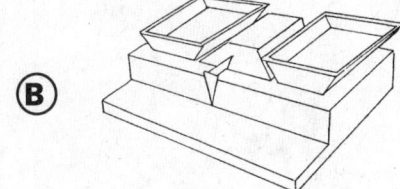
    Ⓑ

    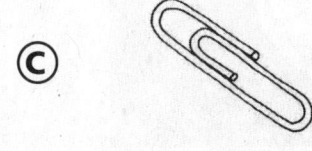
    Ⓒ

    Ⓓ

S4.A.2.2

Use the illustration below to answer question 11.

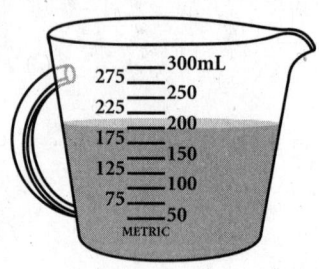

11. What is the volume of the juice in this cup?

  Ⓐ 125 mL

  Ⓑ 150 mL

  Ⓒ 175 mL

  Ⓓ 200 mL

S4.C.1.1

12. Which of the following is an inherited trait?

  Ⓐ a favorite food

  Ⓑ a scar

  Ⓒ dimples

  Ⓓ a chipped tooth

S4.B.2.2

Use the illustration below to answer question 13.

13. What word describes the feature shown?

    Ⓐ valley

    Ⓑ glacier

    Ⓒ lake

    Ⓓ plain

S4.D.1.1

Use the illustration below to answer question 14.

14. Which animal experiences this stage as part of its life cycle?

    Ⓐ lizard

    Ⓑ frog

    Ⓒ grasshopper

    Ⓓ butterfly

S4.A.3.3

15. Some cities have programs to recycle materials such as paper, cans, and plastic. How do these programs help all people?

    Ⓐ by making more resources

    Ⓑ by conserving resources

    Ⓒ by making more pollution

    Ⓓ by increasing the cost of the materials

S4.A.1.1

16. Which object is transparent?

    Ⓐ

    Ⓑ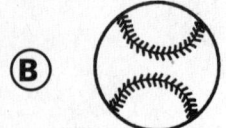

    Ⓒ

    Ⓓ

S4.C.1.1

Use the diagram below to answer question 17.

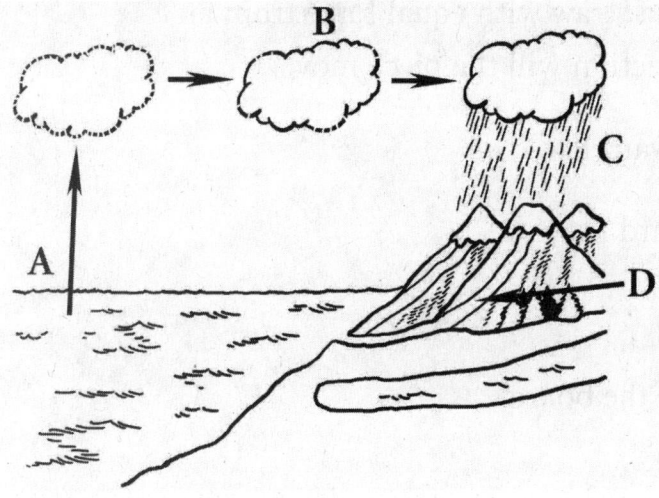

17. What word describes the part of the water cycle that is represented by C?

    Ⓐ condensation

    Ⓑ evaporation

    Ⓒ generation

    Ⓓ precipitation

S4.D.1.3

18. What is one way the environment is changed immediately after a fire?

    Ⓐ Trees grow taller.

    Ⓑ More animals move into the area.

    Ⓒ Habitats are destroyed.

    Ⓓ The number of plants increases.

S4.B.3.2

19. Kelly and Pat sit on opposite sides of a table. They blow on a checkerboard piece through a straw with equal force from an equal distance. In which direction will the piece move?

Ⓐ It will move quickly toward Pat.

Ⓑ It will move slowly toward Kelly.

Ⓒ It will not move.

Ⓓ It will move straight off the board.

S4.C.3.1

20. Which activity affects the environment by increasing air pollution?

S4.A.1.3

Name _____

Date _____

Practice Test 3

Use the illustration below to answer question 21.

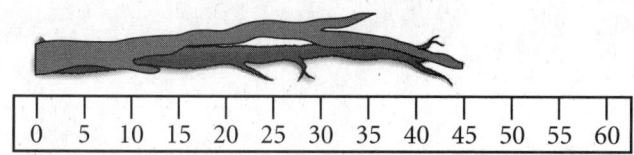

21. What is the length of this branch?

Ⓐ 10 centimeters

Ⓑ 45 centimeters

Ⓒ 75 centimeters

Ⓓ 100 centimeters

S4.A.2.2

Use the bar graph below to answer question 22.

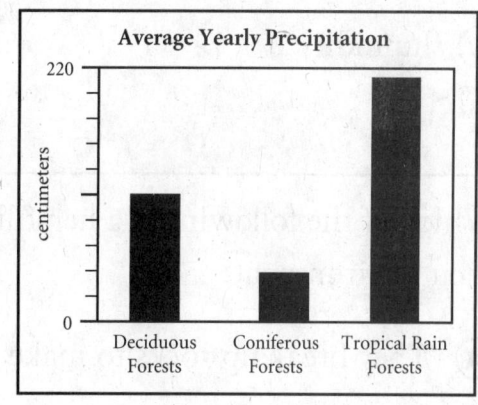

22. What are the two variables shown in the bar graph?

Ⓐ amount of rain and kind of forest

Ⓑ time of year and location of forest

Ⓒ height of trees and time of year

Ⓓ number of trees and location of rain

S4.A.2.1

Use the illustration below to answer question 23.

23. What does this tool measure?

    Ⓐ temperature

    Ⓑ wind speed

    Ⓒ air pressure

    Ⓓ humidity

S4.D.2.1

24. Which of the following is a helpful way that plants affect their environment?

    Ⓐ They break up rocks to make soil.

    Ⓑ They break down dead animals.

    Ⓒ They give off carbon dioxide.

    Ⓓ They form the highest level of the energy pyramid.

S4.B.3.1

Use the illustration below to answer question 25.

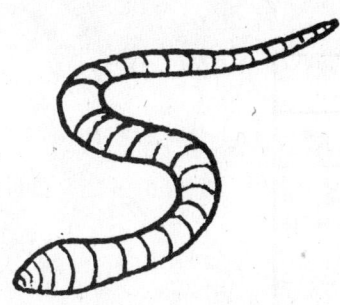

25. What role does this organism play in an ecosystem?

    Ⓐ It is a first-level consumer in a food chain.

    Ⓑ It occupies the highest level of an energy pyramid.

    Ⓒ It helps put nutrients from dead plants back into the soil.

    Ⓓ It converts carbon dioxide and water into sugar and oxygen.

S4.B.3.1

26. Farmers use some crops for food. Some crops, such as cotton, are used to make things. Which of the following is made from cotton crops?

    Ⓐ mirrors

    Ⓑ cereal

    Ⓒ paper

    Ⓓ clothing

S4.B.3.3

Use the table below to answer question 27.

Daily Servings of Food Groups

Food Group	Servings					
bread, cereal, rice, pasta	🥣	🥣	🥣	🥣	🥣	🥣
vegetables	🍴	🍴	🍴			
fruits	🍎	🍎				
meat, poultry, fish, eggs	🥚	🥚				
milk, yogurt, cheese	🧀	🧀				

27. The table shows how much of different types of food you should eat each day. What conclusion can you draw based on the information presented in this table?

Ⓐ Vegetables are healthier to eat than fruit.

Ⓑ You should eat two servings of fruit per day

Ⓒ You should eat more meat than vegetables.

Ⓓ Cheese and poultry have the same nutritional value.

S4.A.2.1

28. Chimpanzees are covered with fur, and they feed their young milk. Which term best describes chimpanzees?

Ⓐ fish

Ⓑ insect

Ⓒ bird

Ⓓ mammal

S4.B.1.1

Name _____

Date _____

Practice Test 3

Use the illustration below to answer question 29.

29. Observe the structures of the plants in the illustration.

   A. How can you use the plant structure to help you determine where these plants live? Explain your reasoning.

   B. In what scientific group of plants are these plants classified?

S4.B.2.1

Name _____

Date _____

Practice Test 3

Use the diagram below to answer question 30.

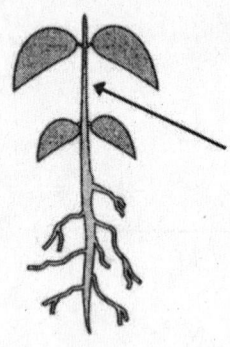

30. Study the picture of the plant.

A. Identify three plant parts. Then identify the plant part to which the arrow is pointing.

B. What is the function of the plant part to which the arrow is pointing?

S4.A.3.1